Jose Luis Ortega Palacios

Smart Home Innovation

Jose Luis Ortega Palacios

Smart Home Innovation

A study in Domotics that offers security and ease of use

ScienciaScripts

Imprint

Any brand names and product names mentioned in this book are subject to trademark, brand or patent protection and are trademarks or registered trademarks of their respective holders. The use of brand names, product names, common names, trade names, product descriptions etc. even without a particular marking in this work is in no way to be construed to mean that such names may be regarded as unrestricted in respect of trademark and brand protection legislation and could thus be used by anyone.

Cover image: www.ingimage.com

This book is a translation from the original published under ISBN 978-613-8-98340-8.

Publisher:
Sciencia Scripts
is a trademark of
Dodo Books Indian Ocean Ltd. and OmniScriptum S.R.L publishing group

120 High Road, East Finchley, London, N2 9ED, United Kingdom
Str. Armeneasca 28/1, office 1, Chisinau MD-2012, Republic of Moldova, Europe
Printed at: see last page
ISBN: 978-620-7-74355-1

TABLE OF CONTENTS

Summary

This research was conducted with the aim of determining how a user can prevent an emergency in which a home automation system system comes to fail and prolong a situation in which it must be attended, here are for that technologists to provide good services and give daily maintenance to prevent any emergency. That is why it is related between automation and intelligent control to housing that allows a very comfortable function as well as good energy savings and good communication between the user and the system, home automation allows to give the knowledge of many electronic devices such as sensors that collects information coming from that is able to capture, process it and then issue an order to the actuators or outputs, allows to give answers to the requirements posed by these social changes in any home design or any building can give a good comfort for the user that helps us in many ways in what is lighting, blinds, water etc..

Key words: home automation, automation, building, electronic devices, energy.

Abstract

Home automation is known to systems that are capable of automating either a dwelling or a building that is too large, which implements many circuits providing energy services, safe and very reliable in what is communication, which are integrated into the interiors and exteriors of a building can be wired or wireless. The objective of the investigation is to determine how a user can prevent any emergency in which the building system fails and a situation in which it must be attended is prolonged, here are the technologists for providing good services and giving daily maintenance to prevent any emergency.That is why it is related between automation and intelligent control of housing that allows a very comfortable function as well as good energy savings and good communication between the user and the system, home automation allows to give knowledge of many electronic devices as is the sensors that collects information that is able to capture, process it and then issue an order to what are actuators or outputs, allows to respond to the requirements posed by these social changes in any house design or any building can be give a good comfort for the user that helps us in many ways in what is lighting, blinds, water etc.

Keywords: Home, automation, automate, building, electronic devices, energy.

Introduction

The objective of the project is 'The protection of the user in home automation and the facilities it offers us', which is part of an automation system in homes and intelligent buildings which began in Ecuador since 2008, In which there was a congress and talked about home automation issues in which it is referred that companies offer system solutions made abroad and here it is only marketed, but the main idea is not to become consumers but manufacturers of equipment for home automation because the knowledge is the same and as Ecuadorians we have a lot of potential in developing technological equipment (Bolton, 2008).

Although the human being is not yet very knowledgeable in the properties offered by home automation is a reality that in a not too distant time loop will be installed in every home. But what is home automation? It is a set of technologies applied in a home in a way that makes it a smart home by saving energy, as having a good security and very good comfort (Millan, 2016).

The word Domotics comes from the union of the words domus (house in Latin) and tica (from automatic), a Greek word for 'that works by itself'.

Domotics has nothing to do with comfort, it also covers many things in different types of areas is also based on the foundation of a good energy saving and how not the protection of the user in their welfare in their good care of an automated home that the user can rest easy. (Morales, 2012).

Home automation has been a good factor in the technological field with the advancement of networks and telecommunications that allows us to navigate and talk about integration at the level of this IP network (Internet Protocol) is very important the electronic part for home automation and smart buildings that now at this time and in the future has facilitated us in many ways in providing us with good security, saving electricity, and multiple services thanks to home automation allows us to innovate in different technological areas.(Gil Moreno, 2017)

It is necessary to have many broader visions in different fields of work all this we can control through a cell phone or laptops of this is based on home automation in a set of techniques oriented to make a home automation or a building that is integrated in much technology in t h e welfare of the user providing good protection and communications through wired or wireless in order to improve and have a good comfort between the user and the home offering new services through a control system (Quintero, 1999).

The main services of home automation are:

Security: Protects the user in many ways in what is smoke sensor in case of fire, personal alarms in what is the domestic use as gas, water and freezer that helps us to visualize that both contain each object and can also turn on an alarm for lack of electricity.

Energy Saving: Intelligently manages energy saving savings in different areas of a home in appliances such as televisions, air conditioners, hot water, irrigation, etc.

Communications: Through a good control and remote monitoring to the house that we can do through a cell phone or a computer that has access to the internet. The home automation installation allows to communicate in different ways such as voice, also including text, images and even sounds with local networks as this local area network (LAN) is known.

Comfort: Making the home more comfortable for the user to enjoy doing different domestic activities such as: opening, closing, turning on, turning off, lighting, air conditioning, blinds, electricity, water, gas, etc. (Laserna, 2014).

The houses have evolved of diverse functions throughout the decades in the social scope as much in the technological area in the decade of the 80's when the premises began to use integrated systems to later be developed as domestic uses in the urban houses and thus this project was implemented worldwide, based on the innovation of the technology that the user had in a very feasible way. (Rybczynski, 2008).

It is evident that we are currently living end times, significant social changes promoted by new technology, these changes are not only noticed in the social and public sphere but also in the private ones is the impact of "modern technologies of transmission and remote communication" which is characterized by the use of technology integration through the phone or through a Pc. (Echeverria, 2016)

The environment in a domotic house makes the user feel satisfied that this is known as a "technological environment" that provide air freshener spaces in house, car, office, etc. It is clear that a domotic house is made for a good comfort in which the person feels more comfortable, the good environment makes the person have a good comfort (Eneo, 2013).

Many are the services that can offer you in many places where one makes these implementations of electronic services but there are few who know the good performance and good comfort that is to have a house of highly technological and that can give you the best for the user who is in good environment and know how to apply this as a daily life.(SAKKAS, 2018)

The advancement of new information and communication technologies (ICT) has become a trend in all areas of life, as it has made that in recent times there has been m u c h talk about intelligent buildings, intelligent areas, intelligent cities and many business opportunities. It is clear that the development of new technologies is related to each other (Clara, 2005).

The world has changed a lot with the technological advancement that has made us see our surroundings in a virtual way in which we focus on every place we are there is technology and how not the smart homes that have evolved a lot in European countries such as Latin America and you can see how far technology has come and will advance. (ALTROCK, 2017)

A set of systems is in charge of regulating and properly managing the electronic elements and household appliances incorporated in a house such as air conditioners, windows, water supply, e t c. (Oberto, 2006).

It gives you a good savings and control of electrical energy facilitating you a low consumption a clear example would be the heating that instead of leaving it on all day you can program for a time loop of minutes, apart another advantage is the security can detect any intruder in your home who wants to steal or a gas leak or a water leak all this plus the comfort to make a high-end home and very comfortable.(Colina, 2019)

An intelligent system makes it very easy for many people to use just with a device that nowadays we call cell phones that connected to a wireless internet network we can do many different types of chores and one being at a distance I could give commands to different devices, we are living in a technological era where there are high levels of automation (Peréz, 2010).

There are many houses which have many benefits one of them is the installation of electronic equipment which allows you to automate the entire area of your home imagine doing all the chores of your home through a cell phone that through the famous internet can do all kinds of daily routines.

In the world we live in, technologically there are different domotic standards to make and conform a digital home, there are different types of standards that allow us to modify and make a house as we want, but the problem is that if we use a standard for a single home, it has to be determined for a single home. (Camps, 2006).

Domotics does not affect anything but rather helps a lot to all kinds of people from children to the elderly because it is to offer a good security service in which we can see how great it is to see a smart home with all the issues and measures that can help us in the near future of much security and to make a home of much productivity. (HOKENSON, 2015)

Many times man has tried to live in a comfortable habitat until the beginning of home automation that has modified his physical environment and that can vary much more in time and that this has evolved in such a short time where there are already houses that are full of many sensors and that can vary many things in the house but use the same devices and are very flexible and adaptable. (Andrade, 2001).

There are disadvantages in smart homes one of these is to invest is very expensive and are very high prices for the user as they are automated systems and also is that you have to do maintenance to these teams which are very complex and expensive and another disadvantage is the interference of data internet speed depending on how many points are connected always connect in a ring (Maya, 2004).

Nowadays there are many domotic systems that are integrated in computer science and many information technologies in which we expect a house with a good comfort, security for our family that is active for any case of emergency and why not a good environment in which the house enables its best services with the help of programming (Graziani, 2018).(Graziani, 2018).

A very important aspect nowadays is that people need to be reliable and safe in the environment in which they operate, whether it is their home or company where they work, and for this reason circuits and systems have been implemented to meet our needs, which makes the user have confidence in the electronic circuits and feel very comfortable. (GERMAN, 2006).

The concept of automation has many years of existence as such, since a person had the idea of connecting two electrical wires to the hands of an alarm clock, so that shortly thereafter, and moved by these hands, the wires would close a circuit consisting of a battery and a lamp. This may have been the moment when the idea of timing an electrical function was born. Later the systems were perfected, primitive at the beginning and much more sophisticated later, until arriving at the present moment where fundamentally the industries base great quantity of phases of production in different types of automatic or timed elements, from the sound of the siren of entrance of the workers or of a student institution, to the preheating of ovens so that when the different workers arrive they find their work stations in optimal conditions (Santiago, 2004). (Santiago, 2004).

In France, which is very fond of adapting its own terms to new disciplines, the word "Domotique" was coined. In fact, in 1988, the Larousse encyclopedia defined the term domotics as follows: "the concept of housing that integrates all the automatisms in terms of security, energy management, communications, etc.". In other words, the objective is to ensure the user of the home an increase in comfort, security, energy savings and communication facilities. A more technical definition of the concept would be: "a set of housing services provided by systems that perform various functions, which can be connected to each other and to internal and external communication networks. This results in significant energy savings, efficient technical management of the dwelling, good communication with the outside world and a high level of security". (Alberto, 1999)

General Objective:

To show how home automation has been updated since the 70's how it has evolved over time and to know how important it is nowadays as we can make good energy savings just by implementing these electronic circuits in homes and buildings, apart from that allows us to have a good security for both the user and the family just by implementing security cameras and electric fences and why not the famous gas leak sensors, water leakage and water expanders in case of an outbreak of fire.

Develop the implementation of home automation to have a good comfort of our home areas such as air conditioning, heating, lighting and others is very important to know how now the new technologies have evolved so much that we can do everything with programming and give orders to our house through a cell phone that is connected to a wireless internet network, is to see the wonder that we can do now in this century known as a technological world.

Specific Objectives:

- To provide good security to the user and to make him feel very comfortable with the structure and design of a smart home.
- Provide guidance to users on home automation that can help us in different areas where we find ourselves through programming.

- Recognize each implementation of the electronic circuits that are installed in a house to offer good comfort, external and internal security.

Theoretical Framework System Description

There are different and varied components that make up the technical management systems and home automation, such as a remote control to manage the installation through the central that manages the entire system in a centralized installation. In this large number of elements that make up an installation, we will begin by addressing the very characteristic and vitally important elements in an automated home, these are the controllers, sensors and actuators, we will describe each of these.

An automated system is able to do any activity as we want to order, it is able to meet all our needs in our home or outside it, it is a real fact that n o w a d a y s we live in European or Asian countries are those that mainly have these technologies and has been developed worldwide to raise smart homes that give us a good security and a pleasant environment and that is why there are now resources to have a good ecosystem based on technology.

- Drivers: A driver is a device or a driver (device driver) is a computer program that allows us to interact between an operating system in this case the computer and a peripheral or hardware that would already be the electronic components which can be schematized as an instruction manual that tells the operating system, how to control and communicate with a particular device.

Nowadays there are as many types of drivers as there are many types of peripherals, it is very common to find more than one possible driver for the same device, but the difference is that each different driver offers a different level of functionality.

Device drivers are programs added to the operating system kernel, initially designed to manage peripherals and special devices. They can be of two types; character-oriented or block-oriented, they constitute the well-known disk drives.

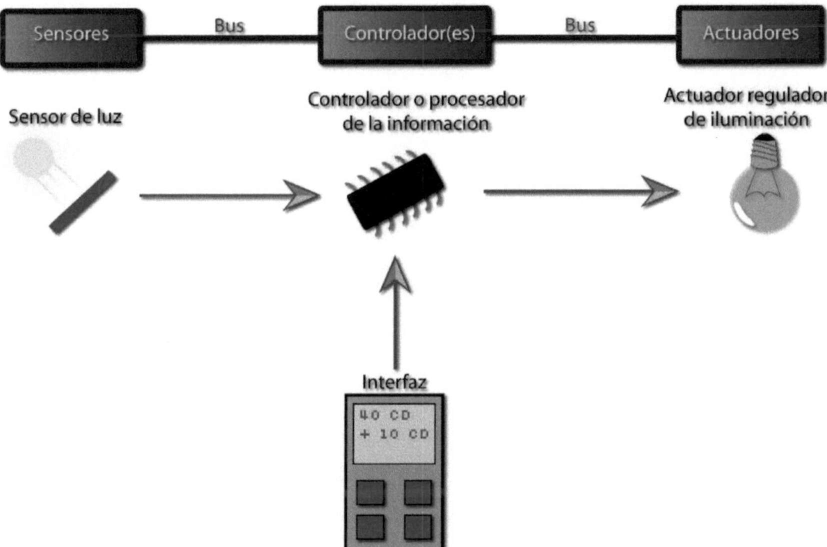

With these controllers of different designs and uses it allows us to have a more feasible daily life while living, it saves us time to do some kind of activity without worrying because everything is under control.

- **Sensors:** Sensors are devices that are used inside the system to evaluate the state of parameters such as ambient temperature, gas or water leakage, etc. The most commonly used sensors are the following:

-       The room thermostat whose function is to register a suitable temperature for the user, where the user is here also has to do with climatic changes and any change to modify the environment according to the system.

-        The indoor temperature sensor whose function is mainly to measure the temperature inside the home.

-        Temperature probes for heating management, the function is to effectively manage the operation of any type of electric heating such as limiting probes for underfloor heating.

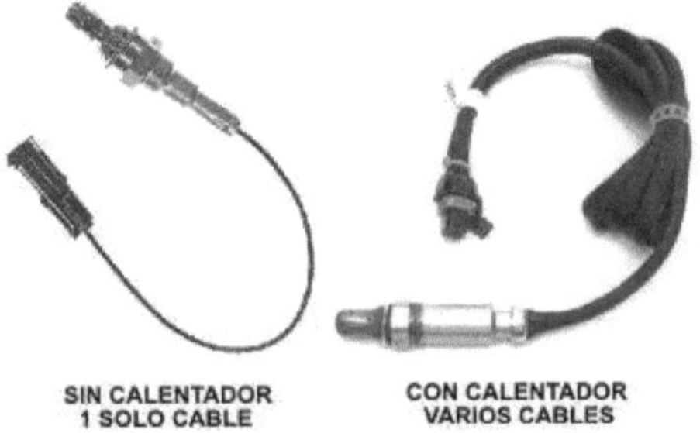

-        Gas leak detector, used to inspect possible anomalies and leaks in home installations either indoors or outdoors.

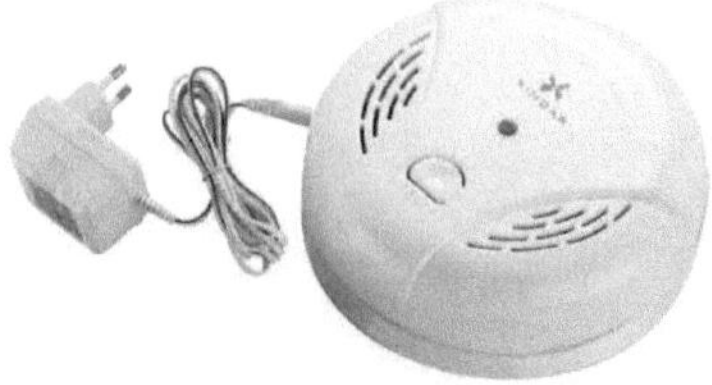

-      Presence sensor, which will try to see or perceive strange movements in the house, for example, intruders or people who are not allowed access to the house.

-      Smoke detector, this sensor is designed to warn us in advance of possible fires that may occur and thus avoid a misfortune.

-      Infrared receiver this fulfills a very important function in a smart home and is that it allows us to give orders to the distance by WI-FI connection and a mobile device.

What controls does it allow us to do? Here come the electrical appliances such as air conditioners, heating, blinds motorized blinds, Tv, radios, microwaves,

 refrigerators, fans etc., all these know how to carry a control to turn on or off or give other orders, but here in home automation we can do all that in a single control that does not provide a great ease and help because it is already registered in the system depending on how we want our settings.

DOMOTICA WiFi – INFRARROJO

El AWIR-1515 está diseñado para controlar cualquier dispositivo electrónico con receptor infrarrojo de 38Khz. Posee una función de aprendizaje de comandos que luego serán enviados por usuario para controlar, por ejemplo su sistema de entretenimiento.

Con una capacidad de almacenar hasta 15 comandos IR, el usuario puede controlar con un mismo módulo WiFi uno o más dispositivos electrónicos.

- The most commonly used actuators are as follows:

What are actuators?

An actuator is a device capable of transforming hydraulic, pneumatic or electrical energy into the activation of a process in order to generate an effect on an automated process.

- Actuating relays, normally mounted on DIN rail, have the function of acting on the elements that need to be actuated, such as air conditioning equipment.

conditioning or a regio system in the garden

DIN rail mounted relays

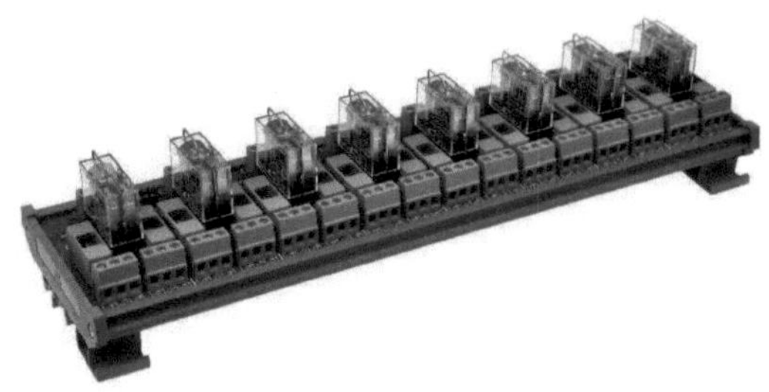

-       Solenoid valves to shut off gas or water supplies will act as a requirement of the system when a humidity sensor stops, as well as in a gas leak detector, shutting off immediately when there is a malfunction in the installation.

\-   Sirens or buzzers (buzzers) that will be activated for cases in which the detention of an alarm by the system is necessary for both acoustic and visual signaling.

\-   The buzzer is an electrical mechanism that emits a buzzing sound as a warning. For its connection you need a two-wire electrical wiring that connects it to an electric push button.
\-   The sound level can reach up to 80 decibels depending on the model, making them a good alternative to the bell chime in noisy environments.

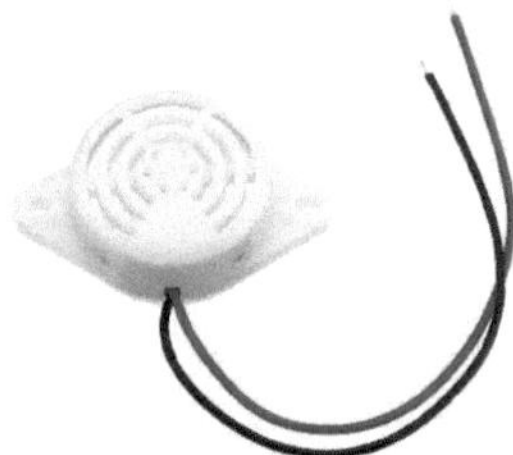

Differences between Domotics and Inmotics

The differences between Domotics and Inmotics are very easy to recognize, and are mainly based on one of great importance of structures. While domotics is about the installation of intelligent systems implemented in common dwellings or family properties which have comfort for the family, immotics refers to the same procedure, but in larger structures such as hotels, work buildings or hospitals. Although this does not mean that the products offered in both branches cannot be used by the other. That is to say, what has presence in a mega construction, could also be adapted to a family home, and vice versa.

The structure of a Domotic house

The smart home is one of the new eras in these centuries but what are the benefits of having or making one of these houses is simple it gives you many advantages such as opening windows, doors or light bulb with your smartphone just having a smart phone and internet you can do many activities inside or outside your home and more in appliances and has ease of having security cameras that while you are in different places you can observe your home, apart from a heating system or cooling system depending on the weather it is very common to find in any house one of these systems, as you can also automate your garage, how does a garage automation work?

This can be located next to your house with a surveillance or alarm system, and how to open a garage door very easy, through an automated system with a control you can open it or with a system that recognizes the plate of your vehicle and automatically this will open, it is also very important energy savings that we can save in a day and how this is possible, it is very easy because each house comes or you can adapt solar panels that are of many benefits that give you these technological houses.

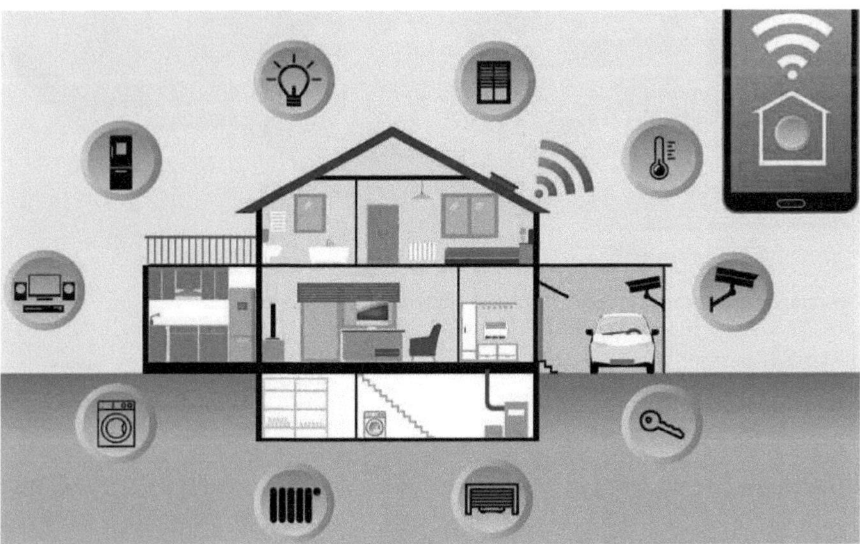

The structure of an Immotic house

Inmotics is a complex large-scale integrated management system, designed for large-scale structures, especially: shopping malls, office buildings, hotels, convention centers, among others.

Building automation can contribute up to 40 percent of the energy savings in a large building. Thanks to it, surveillance systems, elevators, lighting, irrigation systems, heating or air conditioning can be controlled at centralized control points. In this way, the maintenance of the structure and the respective use in each of its areas is optimized to a point close to perfection, a large-scale building.

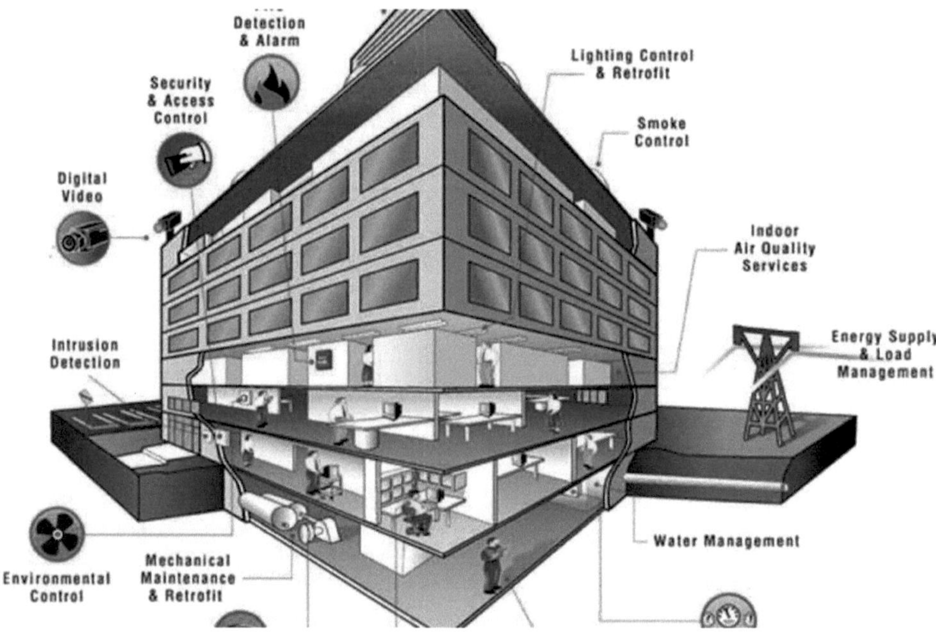

Advantages and Disadvantages of Home Automation Advantages

The advantages that a Domotic system offers us are very complex, let's describe some points:

- Home and Family Protection

Simulating your presence or that of your family when you are not there, through a wireless network you can see your home or give commands through controls for this the house must have mainly camera and alarm systems, the cameras to monitor the home from a distance and the alarms in case of emergencies such as an intruder, accident or fire.

- Adds value to the property

A house with domotic systems benefits a lot today what is quoted at the top of the real estate market and it is also easier to sell one of these houses because it already has all the amenities, incorporates unique features and it is an added value that gives greater category and unquestionably has a trend towards new requirements for modern housing.

- Quality of life

Think of all the routine activities we do every day including entering the house, turning on the lights in the bedroom, bathroom, basement or living room or for example, turning on the appliances or opening the windows, home automation makes all this more comfortable and pleasant by being at a distance with your mobile device or a computer you will do all these things without moving.

- Energy Saving

Adding intelligence to your home, in addition to saving energy, makes your home more environmentally friendly. Everyone knows that solar panels or double glazing save a lot of energy.

It also saves energy because this system monitors the lights and appliances through its sensors, turning them off when they are no longer needed and using them, which saves up to 50% of electricity in a house.

- Investment and useful life of the dwelling

One of the things that most concerns when investing in technology today is its useful life and one of the advantages of home automation systems is that its life is extensive is the new era of housing today and a future with new technological advances.

Disadvantages

The disadvantages of a domotic system are based on the fact that nowadays technology and especially domotics seeks to make our life easier, but does it really always make it easier? These are the disadvantages of home automation systems and we are going to describe some of them.

- No authorized installers

The home automation being something very new in many Latin American countries, is with the disadvantage and almost null existence of authorized installers of home automation, these simply do not exist because the careers in the country do not exist or are very vague, in addition there are no regulatory bodies or institutions of this career, so we find only companies dedicated to this service and therefore with very expensive products and generally do not perform maintenance should be every so often perform proper maintenance to electronic devices so that their life is very long.

- Some degree of complexity

Sometimes those of us who are uncertain in this world forget that all these products are to make life easy for people and not complicate it more, but as some will say "of course that's what it is for", but we must ask ourselves why so many options has that touch screen if all I want to do is turn on the light or lower its intensity, then we wonder why so many remote control, touch screen, iPad, among others to control something that I expected to be made automatic or at least with less search options on what I want.

Advantages and disadvantages of Inmotics

Immotics is based on a larger scale structure and this is where hospitals, shopping malls, hotels etc. come in.

For an owner of a hotel building it can offer a more attractive service while achieving great reductions in energy and operating costs, as well as for the building's users who improve notably in comfort and safety which makes for a much more efficient structure.

The advantages of a control system in buildings and large installations are numerous and we will describe the most important ones:

- Energy savings of 40% to 50% on a daily basis, hotels save full energy in their air conditioning and lighting.
- Savings on maintenance services.
- Personnel management in the building.
- It shows the faults in the structure.
- Increased user comfort and safety.

- Technical alarms.
- Improves worker and building efficiency.

On the other hand, there are also many applications that an Immotic system can offer:

- Lighting control.
- Consumption management.
- Security systems.
- Supervision of the electrical system panel.
- Access control.
- Supervision of the fire system.

The implementation of a control system in a shopping center has 4 main objectives:

- The first and very important objective is the safety and well-being of users in the event of intruders, floods, earthquakes, or gas leaks.
- As a second objective is the complex maintenance of the building, in these implementations is that the control system gives you notice to everything and monitors the electrical system, the ventiloconventor (air conditioning or heating) and the alarm system, thus saving and helping the maintenance service to be all under control.
- The third is energy savings, since the waste of electrical energy is very common in these types of large-scale structures, with good management of lighting and air conditioning a large amount of energy can be saved, amortizing the annual investment.
- As a last objective is to help and provide a good management by the complex and is a great advantage is the ease of management and staff savings that report a control system.

The advantage of implementing a control system and that management complies in a Spa or Fitness Center are the following parameters:

- Air conditioning.
- Air quality control.
- Water temperature.
- Lighting.
- Access and billing controls.

All these allow the creation in function that you are going to give to these types of places.

Advantages in the educational areas there are many fundamental parameters to ensure the welfare of students and teachers and the proper functioning that would have the classrooms or the educational establishment are:

- Air conditioning.
- Lighting.
- Rooms of good comfort and pleasant atmosphere.
- Security.
- Alarm systems.

Advantages in hotels, one of the main sources of income in each country and their owners invest in these luxurious automations for the guest to feel very well attended and marveled at the good functions that bring and good service we will name the following points that a hotel brings in its control system:

- Electrical panels.
- Lighting control.
- Alarms in the hotel.
- Consumption measurement.
- Control of air conditioning and ventilation systems.

A control system can be placed in the hotel rooms to allow the following activities:

- Access control.
- Presence sensor.
- Television system control.
- Climate control.
- Control of services (cleaning).

Disadvantages in Inmotics

The disadvantages are very few and that is because this system does not suffer from many flaws when implemented, it is very secure and very complex and easy to use.

The main disadvantage of an inmotic system for the personnel of any productive sector: Hotel, Industry or business is that this would lower countless sources of work if any of these implement these automated systems.

Its cost is another disadvantage to automate a large-scale building automation system it takes a lot of money to invest in acquiring the right building or company, but over time you will recover your investment.

People who do not have much knowledge about this, could sometimes affect the system of the building in what is electrical system, for this should be a person who has a little training on how to use the controls and what to do in case of a malfunction.

Systems Topology

The term topology refers to the way in which a network is designed along the connection lines, either physically (based on certain characteristics of its hardware) or logically (based on its software).

The topology of a network is the representation of the relationship between the devices (usually called nodes), and all the links that connect them to each other, the types of topology are divided into:

- Star topology.
- Ring topology.
- Bus topology.
- Mesh topology.
- Tree topology.
- Mixed topology.

Star Topology

The Star Topology is a network that is directly connected to a central point and all communications will be done through a node, the devices cannot be directly connected to each other, therefore not much information traffic is allowed.

Feature

It has an active central node that normally has the means to prevent echo-related problems and is used for LAN local area networks.

What is it used for?

It is mostly used in LAN networks also known by its acronym as local area networks that have a router, and a switch or hub following this topology.

Star netting materials

-        Category 5e or 6e UTP cable.

-        RJ45 connectors.

-        Gutter.

-        Switch.

-        Server.

-        Communication protocols.

-        UTP cable (One twisted pair).

Topology Ring

A ring network is one in which each station has only one incoming and one outgoing connection, each station has a transmitter and a receiver that acts as a translator passing to the next station.

Advantages

-        The system provides equal access for all computers.

-        Ease of data flow.

-        Performance does not drop when many people use the same network.

Disadvantages

-        The channel will usually degrade as the network grows.

-        This topology is slower than the others, because the information must pass through all
the intermediate stations before reaching its destination.

-        Difficult to diagnose and repair problems.

Features

-        The cable forms a closed loop forming a ring.

-        All computers that are part of this network are connected in a ring.

-        Ring networks commonly use the (Witness Pass-through) model as an access method.

Bus Topology

A bus network is characterized by having a single communications channel, also called bus, trunk or backbone, to which the different devices are connected, i.e. they all share the same channel for good communication.

Construction

The ends of the cable are terminated with a coupling resistor called terminator, which in addition to indicating that there are no more computers at the end, allows the bus to be closed by means of an impedance coupling.

Advantages

-        Ease of implementation of this topology.
-        Simplicity in architecture.
-        It is a space-saving network.

Disadvantages

-        There are limits on the equipment depending on the signal quality.
-        Performance decreases as the network grows.
-        High transmission losses due to collisions between information.

Mesh Topology

The mesh topology is a type of network that are interconnected in the network devices and computers, thus facilitating the allocation of most transmissions, even when this fall any connection, ie it is a network configuration where all nodes cooperate with each other to receive and transmit information, this topology is used more by wireless networks.

Features

To determine the routes and ensure that they can be used, the network must be self-configuring and connected at all times.

Since there is a lot of physical addressing (MAC) data flowing in the network to establish this route, the Maya topology may be less efficient than the star network.

Advantages

- Resilient to problems, this generates great redundancy, which serves to keep the network active even when problems are noticed.
- There are no traffic problems since it handles a large amount of information and multiple devices can be connected to each other, without collapsing the network and provides high security and privacy.
- Easy scalability, meaning that each node acts as a router, therefore, no additional routers are required, the network size can be changed easily and quickly.

Disadvantages

- The initial setup is one of the main disadvantage, as implementing this requires a   lot of time when configuring it.
- Increased workload, each device has a lot of responsibility, the device not only has to serve as a router, but also has to send data.
- It is very expensive to implement a large-scale network because it requires a large number of cables and ports for both input and output communication.

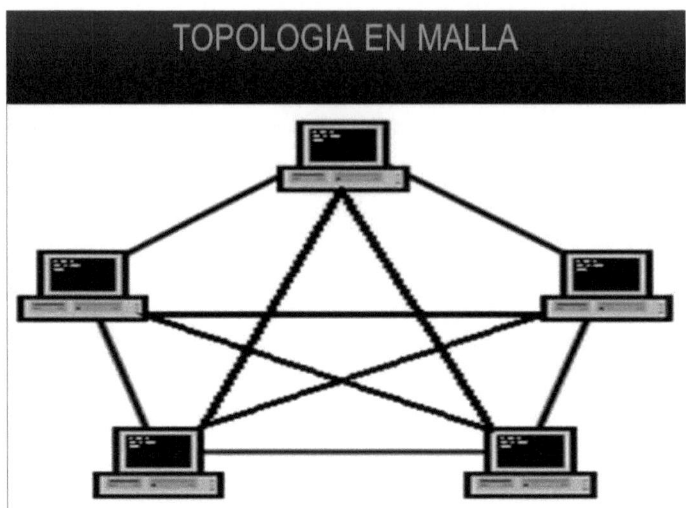

Tree Topology

The tree network is the combination between the bus topology and the star topology which provides more servers to the network, which is called a hierarchical topology, this topology only has a central node in which the other devices are connected to build a single hierarchy, and only applies this topology when the network is large and it is not recommended to use this topology for a small network because you would have to use more cables generating a large waste.

Features

-        A point-to-point connection, each computer has a direct connection to a hub and also each part of the network is connected to a main cable.

-        It has a wider hierarchical relationship than the other topologies and has a minimum of 3 levels, which helps to distribute a large amount of information to the network.

-        The use of the tree topology is mainly used in a network that covers a wide area, it is very feasible if the workstations are located in different clusters.

Advantages

-        This topology reduces network traffic.

-        It is highly compatible for various systems in both hardware and software.

-        Devices connected in another network hierarchy are not harmed if any of the devices in one of the network branches are damaged.

-        Easy detection of errors is very easy to find and the network administrator can correct them in an instant.

-        The access to the computers, due to the tree topology, is for a very large network and any computer will be able to have better access to any device in the network.

Disadvantages

-        A large amount of cabling is required compared to bus and star topology.

-        The implementation of this network is very costly.

-        As the network grows, more nodes are added, making maintenance more difficult.

-        The single point of failure, if the backbone cable of the entire network breaks, both parts of the network will not be able to communicate with each other.

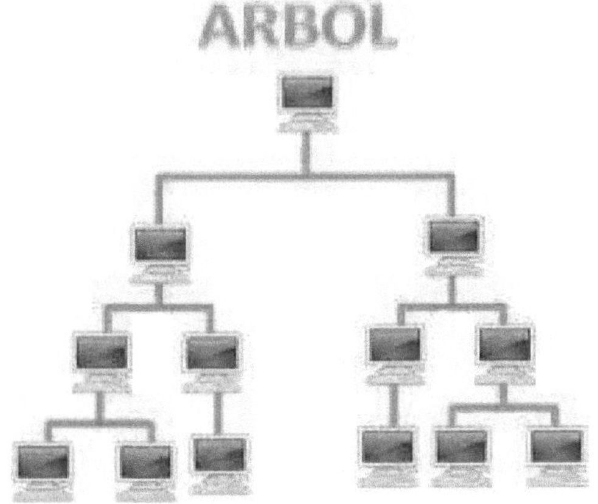

Mixed Topology

This type of network topology is one that uses between two or more different network topologies, it has as a combination of bus topology, mesh topology, ring topology and star topology.

The topology determines how a network will be built, it contains the design of the configuration with the link and the nodes to relate to each other, this is for the network to have a good performance and to organize in the best way to consider the size of the installation and the money available.

Features

It is a scalable topology that can be easily expanded and is very reliable, but at the same time it is a very expensive topology.

A mixed topology occurs only when two or more topologies are joined together, for example, if a star topology is connected as another topology this makes the network faster and more productive.

Mixed topologies exist mostly in enterprises or high-end buildings where each area has its own networked technology.

Advantages

-       This topology is very flexible, reliable and has higher fault tolerances.
-       It is able to utilize the strongest aspects of other networks.
-       Easy in troubleshooting, they are very easy to diagnose and correct, because the connection points on the network hubs are very close together compared to the total size of the network.
-       Easy network growth, scalable as other computers with different topologies can be connected to existing networks and modular in construction, allowing easy integration of new hardware components, such as additional connection points.

Disadvantages

-        Installation is difficult and the design is complex, so maintenance is high and very costly.

-        When implementing a mixed topology, the monetary cost must be considered, including the requirement for high-end equipment.

-        A lot of cabling, so a lot of redundancy in cabling and backup r i n g s are needed to guarantee the reliability standards of the network.

Online applications used to automate a house

An automated home requires having everything under control, in fact, is the main reason for smart homes and it can be done from the home automation central or from our mobile device and it is a great advantage of smart phones that we carry it everywhere and we all have one.

"Eugene Polley invented in 1955 the first remote control for television."
Another important point is that, on the continent of Europe, there are very important markets through the internet to install Domotics applications on our IPad or IPhone that cost around the 40.00 euros, but it offers us almost infinite possibilities to do from our cell phone.

There are programs that you can even get for free, we just have to know which one is better for our comfort and which ones are more related to our environment and home, we are going to name

31

3 totally free applications.

WINK: It is an app with which you can control up to 100 devices without your system collapsing and it also includes an alarm for domestic catastrophes such as the dreaded water leaks.

Imperihome: According to experts, this is one of the most downloaded applications to automate a home, until recently only Android (cell phones) users could use it and now it can be downloaded for IOS (computers).

Presence: This application has been the inverse of the previous one in that it can only be used through laptops but the strength of the application is the control of the house through security camera systems is a strength that makes it very interesting.

Means of transmission

The means of transmission of information, interconnection and control between the various devices of home automation systems can be of various types. The main means of transmission are:

- Proprietary Wiring - Proprietary wiring is the most common means of transmission for home automation systems, mainly of the shielded pair, twisted pair (1 to 4 pairs), coaxial or fiber optic type.
- Shared Cabling - Several solutions use shared cables and/or existing networks for the transmission of your information, for example, the power grid (carrier currents), the telephone network or the data network.
- Wireless - Many home automation systems use wireless transmission solutions between devices, mainly radio frequency or infrared technologies.

When the transmission medium is used to transmit information between devices with the function of "controller" it is also called "Bus". The bus is also often used to power the devices connected to it (e.g. European Instalation Bus - EIB).

Materials and methods

The investigative work began to be carried out on November 5, 2019 and concluded on March 08, 2020, whose project was investigated in the Instituto Superior Tecnológico Portoviejo as well as in areas that have implemented these electronic circuits such as: State buildings, hospitals, buildings of the emergency committee at the National Level here enters; the Fire Department, ECU 911, National Police, etc.

In the materials it is based on the Arduino that a microcontroller that this has been done in different houses to be able to automate them and give commands to sensors, lights, doors, windows and automated rooms.

A survey was also developed for both the population and university students to have as objectives how important is home automation both in the city and nationally and how it can facilitate or security can give us as a result to implement electronic circuits in our homes or buildings, It is very clear that home automation goes beyond a smart home because it facilitates different activities while we are not inside it, each person who filled out this survey liked the idea of implementing sensors, security cameras and lights that with a single movement were turned on, this survey was conducted in order to see the perspectives that people have when hearing about a home automation system.

DIAGNOSIS

The field research is based on surveys conducted in the city of Portoviejo to gather information to analyze the way and the sector in which home automation can be implemented, for the processing of the information a logical-qualitative analysis was performed, since through this the most relevant points of t h e research are exposed and also a logical-quantitative analysis is performed since through statistical methods such as tables and graphs will be given greater facilities to evaluate the variables and their subsequent analysis according to the results obtained.

This survey reflected the ideas as users plan to have a fully automated home and we will shortly show the results of each question asked in the survey.

Results

Order	Alternative	Frequency	Percent
A	Always	20	50,00
B	Sometimes	10	25,00
C	Rarely	5	12,50
D	Never	5	12,50
Total		40	100

Table 1; Have you visited a domotic house?
Source: Universidad Técnica De Manabí

A smart home causes people a great impact on what is a good automation and also as protection towards the person in what reflects the table 1, most of the people surveyed have seen or visit a house or building in a smart way or is a very effective way for this to evolve much more in the future in what creates a great impression on others, it should be emphasized that these implementations serve a lot in any house that are well structured as long as there is a good comfort and efficiency for the user for protection in any emergency.

Order	Alternative	Frequency	Percent
A	All	10	25,00
B	Some	20	50,00
C	None	10	25,00
Total		40	100

Table 2; Do all the buildings you have visited or know of have good parking?
Source: Universidad Técnica De Manabí

In every smart home or house there should be a good parking lot to have more space and conformity as a good signaling in case of emergency, the results reflected in Table 2 is that the people who conducted the survey many people gave opinions that it would be very useful to have a good park which is programmed in order that the user just get home the parking lot is opening which will save time and have a good security, since there are few who have a system like this.

Order	Alternative	Frequency	Percent
A	Much	30	75,00
B	Little	5	12,50
C	Nothing	5	12,50
Total		40	100

Table 3; Do you agree with the implementation of electronic circuits in homes?
Source: Universidad Técnica De Manabí

An implementation of electronic circuits in a house is a very good option because it facilitates us a lot being a pc or a smart cell phone which allows us to do many domestic activities, in table 3 reflects that many people agree with this great implementation that gives us ease to different types of activities in a smart home or building that through this gives us security, comfort and energy savings.

Order	Alternative	Frequency	Percent
A	Much	13	32.5
B	Little	20	50,00
C	Nothing	7	17,50
Total		40	100

Table 4; Are you very interested in the issues of a smart building in the city?
Source: Universidad Técnica De Manabí

Intelligent buildings are very important because they are designed and structured in a very safe way in all types of emergencies are sensorized either in their doors, corridors, and in different areas against fire, earthquakes and theft, in table 5 the people who conducted the survey said that it is good to implement these circuits in all buildings and not only the government, this automated system will help protect the person from all dangers is important to the welfare of one with a good compliance and that the user relates to the electronic things that is innovative worldwide.

Order	Alternative	Frequency	Percent
A	Much	20	50,00
B	Little	13	32,50
C	Nothing	7	17,50
Total		40	100

Table 5; Are there many buildings or houses in your city in a smart way?
Source: Universidad Técnica De Manabí

The buildings or domotic houses have been of much use in which the user can realize that technology advances day by day with the help of an electronic circuit one can see beyond what can be done in Table 6 as results does not give to see that there are varieties of houses that use at least 3 or 4 circuits intelligently already using a security camera or an electric fence or why not an electronic garage is already a house intelligently.

Order	Alternative	Frequency	Percent
A	Always	12	30,00
B	Sometimes	8	20,00
C	Rarely	9	22,50
D	Never	11	27,50
Total		40	100

Table 6; Have you seen these features with sensors and cameras in the houses?
Source: Universidad Técnica De Manabí

In certain houses and buildings there are these sensor systems but in most there are these systems which provides security and compliance to the user in that form security because by implementing these cameras are activated with an alarm and anything that puts the user at risk by a mobile can activate it and what is sensor either have a door that with the movement of a person opens and why not the windows that has its electric shutter, in table 7 many people have seen how others have not, buildings or houses with these features, but if they gave many opinions that it is good to implement this in every house.

Order	Alternative	Frequency	Percent
A	Excellent	28	70,00
B	Very good	10	25,00
C	Good	1	2,50
D	Regular	1	2,50
E	Deficient	0	0,00
Total		40	100

Table 7Do you agree with this implementation in your city to provide good security?
Source: Technical University of Manabí

This implementation on the subject of Domotics is a very broad factor, but of high efficiency and above all of very little demand since in certain cities do not have much knowledge of this, but it is good for people to know about this, table 7 most people agreed to this to have a safe city and smart way that allows us to do many activities daily routines while we have a safe city.

A fully smart home with many electronic devices allows us to m a k e our life easier because we would be the ones to do the command through a smart cell phone, Table 8 reflects as results that most would like a smart home because of its security, compliance, energy savings, and because it is with wireless networks.

Order	Alternative	Frequency	Percent
A	Excellent	30	75,00%
B	Very good	6	15,00%
C	Good	3	7,50%

D	Regular	1	2,50%
E	Deficient	0	0,00%
Total	40	100	

Table 8; What would it be like for you to have a smart home?
Source: Universidad Técnica De Manabí

The survey for many people found it very interesting for people to implement electronic circuits in the structures whether smart homes or buildings as this has evolved so intelligently that even in the houses we can see not many electronic devices, but if most implement them not completely, but if little by little by the issue that is very expensive.

Discussion

In home automation there are many factors in which there are advantages and disadvantages, one of the advantages in smart homes is that they offer a good comfort and have very safe systems, but one of the disadvantages is that they are very expensive to have a complete implementation in the house as there are many programmed systems and require both a good internet and components that are expensive and it should be noted that there are different types of circuits.

The new knowledge of the project is that there are parts in which the home automation are integrated circuits of other levels that are more extensive and this sometimes fail are very complicated to repair them in the results emphasizes that it is of great help a home automation in all areas the only bad thing would be the price to acquire and make one of these, but to offer you the best services that can give you a house.

Conclusions

It is necessary the implementation of electronic circuits in homes and building as they now come automated and programmed for the user that one feels safe to live because it is surrounded by no number of sensors such as motion sensors, light bulbs, doors and air conditioners that are programmed for a certain time limit to make a cooling is necessary that users are aware of the good security that could help us in different areas while one is away from home by means of a certain time limit to make a cooling, doors and air conditioners that are programmed for a certain time limit to make a cooling **is** necessary that users are aware of the good security that could help us in different areas while one is away from home through a cell phone can observe what is happening inside the house is necessary to recognize that all these circuits work connected in a wireless network and are already programmed for a period of time. programmed for a period of time are very effective because all this implemented in homes offer good energy savings in which we greatly facilitates the costs and a house is made up of a good anti-seismic structure and in any emergency case of any gas leak these sensors will detect it and have its function and also at night all the doors are activated locks and even with voice recognition one can give a signal.

Bibliographies

AlLTROCK, J. A. (2009-2010). Domotics and Inmotics. Retrieved from Domotics and Inmotics: http://www.nebrija.es/~jmaestro/ATA018/Domotica.pdf.

Andrade, M. (2001). Department of Building Technology. Madrid. Automated simulation of architectural lighting in facades. Faculty of Architecture and Design. University of Zulia.

Bolton, W., & Ramirez, F. J. R. (2001). *Control engineering* (pp. 1-3). Mexico: Alfaomega.

Camps, R. D., Yagüe, J. L. P., Luján, J. L. P., Blasco, P. P., & Escribá, J. C. C. (2006). SCHome architecture: Remote access to the digital home. *Proceedings of the XXVII Jornadas de Automática (JA2006)*, 348-354.

Clara (2005) LSB Smart Technology. Retrieved on 10/04/2016, from X1O System How It Works: http://www.lsb.es/imagenes/x10_introduccion.pdf

Colina Mendoza & Cristóbal Romero Morales, F. J. (2007). Domotics and Inmotics. Intelligent homes and buildings. 2 Edición. Mexico: Alfaomega Grupo Editor. Cubero, A. C. (09 2008). University of Seville Library. Retrieved from Biblioteca Universidad de Sevilla

Echeverría, J. (1995). *Cosmopolitas domésticos*. Barcelona: Anagrama. Heidelberg, pp. 227-246.

Éneo Ramírez. (2007). Mushuc Runa building for November in the city of Quito. La Hora. Retrieved from http://www.lahora.com.ec/index.php/noticias/show/601253/-

Graziani G, 2018, Prototype Description of a Bioclimatic Housing With Domotic Approaches Applying VRML, Vol. 1 pp 2-3.

Lasserna (2014). SENSOR NETWORK. 2003, from ARCHITECTURES PROTOCOLS Website: https://bibdigital.epn.edu.ec/bitstream/15000/938/1/CD-1838%282009-01-21-12-25-09%29.pdf

Luis Hokenson Álvarez, A. M. (June 16, 2007). UNIVERDSIDAD DE OVIEDO. Retrieved from REPOSITORIO UNIVERDSIDAD DE OVIEDO.

Maya, 2004 Karimanal, A. E. (02 2012). Spanish Journal of Electronics. Retrieved from Revista Española de Electrónica: http://www.redeweb.com/_txt/687/p56.pdf Kemisa. (n.d.). kemisa. Retrieved October 03, 2017, from kemisa: http://www.kemisa.es/es/circuitos-para- ordenador/42-encendidoautomatico-del-pc.html.

Millan & Carlos (2016). LEARN QUICKLY TO PROGRAM. 2006, from DOMÓTICA, Vol. 1 pp 3-4.

Morales Edison, F., González, V. M., Poo, R., García, M., & Olaiz, R. (2010). Design and Development of An Automatic Small-Scale House For Teaching Domotics. Proceedings of the Frontiers in Education Conference, 2010. On 31st Annual - Volume 01. Washington, DC, USA: IEEE Computer Society, pp T3c-1-5.

Moreno Gil, J., Lasso Tarraga, D., & Rodríguez Dieguez, E. (2017). *Automated installations in homes and buildings (*1st ed.). Madrid: Paraninfo Cengage Learning.

Oberto Palacios & Manuel Cabello, M. S. (2016). Basic electrical circuits II (Interior electrical installations). Editex.

Perez David. Domótica Viva (November 08, 2012). Domótica Viva. Retrieved from Domótica Viva: http://www.domoticaviva.com/X-10/X-10.htm Flores, J. A. (March 2007). Repositorio Escuela Politécnica Nacional. Retrieved from Repositorio Escuela Politécnica Nacional: http://bibdigital.epn.edu.ec/bitstream/15000/391/1/CD-0798.pdf.

Quintero González, J. M., Lamas Graziani, J., & Sandoval Gonzáles, J. D. (1999). *Control systems for homes and buildings: domotics.*

Rybczynski, W. (2008). *Automated installations in homes and buildings (*2nd ed.). Buenos Aires: Emece Editores.

Sakkas (May 22, 2018). Smart buildings take over Quito. La Hora. Retrieved from http://www.lahora.com.ec/index.php/noticias/show/1038174/1/Edificios_inteligentes_se_toman_Quito.html#.VDtOT_l5N7k LSB. (n.d.).

VILLALBA MADRID GERMAN (2006) INTRODUCTION TO HOME AUTOMATION. INTELLIGENT BUILDINGS

http://www.cib.espol.edu.ec/Digipath/D_Tesis_PDF/D-35310.pdf

FERNANDEZ Valentín, El Hogar Digital Creaciones Copyright, 2005 PEREZ Alberto,

Introducción a la Domótica, Prentice May, 1999.

UK Essays (2019). Tree Topology: Advantages and Disadvantages. Retrieved from: ukessays.com.

Studytonight (2019). Types of Network Topology. Retrieved from: studytonight.com.

Junaid Rehman (2019). What is tree topology with example. IT Release. Retrieved from: itrelease.com.

Printed by Books on Demand GmbH, Norderstedt / Germany